Pic Puzzles

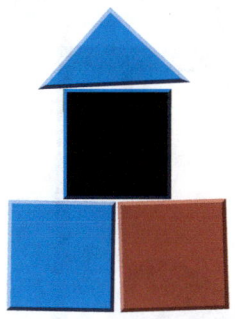

by Melvin Jefferson
illustrated by Julia Gorton

HMH

Copyright © by Houghton Mifflin Harcourt Publishing Company

All rights reserved. No part of this work may be reproduced or transmitted in any form or by any means, electronic or mechanical, including photocopying or recording, or by any information storage and retrieval system, without the prior written permission of the copyright owner unless such copying is expressly permitted by federal copyright law. Requests for permission to make copies of any part of the work should be submitted through our Permissions website at https://customercare.hmhco.com/contactus/Permissions.html or mailed to Houghton Mifflin Harcourt Publishing Company, Attn: Intellectual Property Licensing, 9400 Southpark Center Loop, Orlando, Florida 32819-8647.

Printed in the U.S.A.

ISBN 978-1-328-77229-9

4 5 6 7 8 9 10 2562 25 24 23 22 21

4500844736   A B C D E F G

If you have received these materials as examination copies free of charge, Houghton Mifflin Harcourt Publishing Company retains title to the materials and they may not be resold. Resale of examination copies is strictly prohibited.

Possession of this publication in print format does not entitle users to convert this publication, or any portion of it, into electronic format.

My brother and I play a game with shape pieces. We like to see who can make a picture using the most pieces.

2   How many pieces do you see?

I take 4 pieces. What can I make?
I know, I'll make a boat.

10 – 4   How many pieces are left?   3

I put my pieces back.
It's my brother's turn.

4   4 + 6   How many pieces are there in all?

My brother takes 6 pieces.
He makes a tall tower.

10 − 6   How many pieces are left now?

My brother puts his pieces back.
I take 4 pieces and make a castle.

6   10 − 4   How many pieces are left?

My brother takes 6 pieces and adds them on to mine. We make a rocket ship using all the pieces.

4 + 6   How many pieces did they use?

Responding

Problem Solving

Blast Off!

Draw

Look at page 7. Draw the rocket ship you see.

Tell About

Draw Conclusions Look at page 7. Tell about the different shapes that make up the rocket ship.

Write

Look at page 7. Write the number of shapes the children used to make the rocket ship.